D1037548

ST. STEPHEN'S EPISCOPAL CHURCH
82 KIMBERLY DRIVE
DURHAM, N. C. 27707

# Sister Earth

# SISTER EARTH

Ecology and the Spirit
Helder Camara

New City
London   Dublin   Edinburgh

First published under the title of *Quem Não Precisa de
Consversão?* in 1987 © Edicoes Paulinas, Brazil

First published in Great Britain 1990
by New City
57 Twyford Avenue, London W3 9PZ

© English language edition New City – London

Translated and adapted for English by *Connie Murphy*
and *Callan Slipper*

Illustrations and cover design by *Duncan Harper*


British Cataloguing in Publication Data

Camara, Helder
  Sister earth : ecology and the spirit
  1. Christian life
  I. Title   II. Quem nao precisa de conversao. *English*
  248.4

  ISBN 0-904287-33-5

Typeset in Great Britain by Chippendale Type,
Otley, West Yorkshire

Printed in Great Britain by Dotesios Printers Ltd,
Trowbridge, Wiltshire

ST. STEPHEN'S EPISCOPAL CHURCH
82 KIMBERLY DRIVE
DURHAM, N. C. 27707

# Contents

## II The School of the Universe

# Introduction
# A Man of Peace

"When I feed the poor, they call me a saint; when I ask why the poor have no food, they call me a communist," Dom Helder Camara once said. Indeed, he has become renowned all over the world for his defence of the poor, even in the face of military dictatorship.

A peaceful revolution carried out by social action is at the heart of Dom Helder's life and work. He is not a man to encourage violence or bloodshed, but he urgently seeks justice for the oppressed, denouncing injustice and piloting new projects to help those in need. Hence it is not surprising that ecological issues greatly concern him as well. For yet another instance of oppression can be seen in the destruction of the environment, as greedy or thoughtless people destroy what belongs to all.

Dom Helder's answer for the problems that threaten our planet is the same as his reply to the issues of poverty and human exploitation of others. It is a call to the values of the Gospel, a challenge to live in a fully Christian way. Only thus can human limitations be faced realistically; only thus can human beings find the motivating force to overcome them.

The present book, therefore, has grown like everything else about Dom Helder from his friendship with God. It has deep spiritual roots.

These roots were first fostered by his family, a poor one living in Fortaleza, in the Brazilian state of Ceara. He was born in 1909 on 7th February to Adelaide Pessoa Camara, a school teacher, and Joao Camara Filho, a bookkeeping clerk. Adelaide, his mother, wanted to call him José, but eventually she gave in to his father's wish to name the child after the small Dutch port of Den Helder. The young Helder was one of thirteen children, and was among the eight survivors of an epidemic that killed five of his brothers and sisters.

At the age of eight he received his first Holy Communion, on 19th September 1917, and just six years later, he went to study for the priesthood in the Archdiocesan Seminary of Fortaleza. His father said to him some words about the priesthood which have always stayed in his mind: "Being a priest and thinking of yourself can never go together. The priest has to consume himself in working for others."

In 1931, when he was twenty-two years old, he was ordained. At the same time he began his public life, becoming planning director for education in the state of Ceara. Then, when he was twenty-seven, he moved to Rio de Janeiro where he remained for a further twenty-eight years. In Rio he became one of the directors of the Department of Education and Culture. During the same period, from 1946 to 1962, he was very active in Brazilian Catholic Action and became its national assistant.

At Rio he was ordained bishop in April 1952. His friend Dom José Tavora suggested he should take as the motto for his episcopate: *In manus tuas* ('in your hands'). This is his characteristic way of living; he puts everything into God's hands, trusting in God's Providence.

From 1952 to 1964 he was a member of the Federal Council of Education, the Supreme Council for Immigration, as well as the director of the Episcopal Conference of Latin America. Dom Helder, furthermore, was the moving spirit behind the founding of the National Brazilian Bishops' Conference in 1952. He was its first general secretary, and it was during his time in office that he attended the Second Vatican Council (1962–1965).

Dom Helder's concern for the poor led to his attempts to solve some of the problems in the slum area of Rio de Janeiro. This area, called the 'favelas', was in drastic need of proper housing, so in 1956 he began a programme to provide places fit to live in. This was followed in 1959 by the 'Providence Bank' which was set up to give financial support to people with economic difficulties and to assist in welfare projects. At the same time a trade fair (called the 'Providence Fair') was started to acquire funds for the bank's activities.

Then in 1965, having left the post of general secretary, he became the National Brazilian Bishop's Conference's director for Social Action.

His intense social activity had already been given a boost when in 1964, Dom Helder became Archbishop of Olinda and Recife, Pernambuco. In this position he continued to mount a peaceful, non-violent struggle for the benefit of the poor. This national and international action against injustice has been recognized in the award of many prizes and honorary titles. Even in the worst days of military dictatorship his voice was always heard, speaking out on behalf of the oppressed.

Today he is Archbishop Emeritus of his diocese, a man of faith and simple prayer, who once said: "I don't give myself up to penance . . . I don't scourge my flesh. Instead, I wake up every day at 2 o'clock in the morning to keep my vigil. My vigil consists in talking with God, the speech of old friends which has no formality."

It is from these hours of vigil, spent in friendly conversation with God, that Dom Helder's many books emerge. *Sister Earth* is one beautiful example of them. But despite its simplicity, in order to catch the full meaning of this book, we shall have to have something of the same spirit as Dom Helder, something of the same concern that motivates him. And we shall have to have something of his humility and poetic sensitivity.

# I
# The School of God

# Humility in Creation

# GOD KNEW WITHOUT A SHADOW OF DOUBT

God knew it would have been impossible to create another God: another Infinite Wisdom, another Infinite Sanctity, another Infinite Creative Power . . .

In creating he would, necessarily, be creating the imperfect, the finite, the limited.

With a humility that moves us to the very depths, and with an audacity that could only come from God, he created!

What humility, for Supreme Perfection to create the imperfect!

## The Artist

Every artist hopes one day to reach perfection. When a new work is begun – a new creation in music, literature, painting or sculpture – the artist is convinced that this time perfection will be attained. But as the work develops, the doubts start to come:

'No, not yet, not this time, it seems.' When the work is finished, the gap between dream and reality makes the artist want to destroy the little monstrosity!

## Kant's Scandal

In fact Kant, the great German philosopher, when he meditated upon Creation, even came to doubt the existence of God.

# THE HEIGHT OF DIVINE HUMILITY

The height of divine humility in the creation was when God, from all his creation, chose the human creature and made it co-creator to complete the creation and to help nature express its full potential.

No doubt all other creatures, as well as being amazed and perhaps envious, were very curious to find out to what extent humans would be able to respond to such an astonishing divine plan.

If we look at human intelligence, it is easy to see how humans have helped to complete the creation initiated by God. This is evident not only when we consider the discovery of nuclear energy and flights into space, but also in the discovery of fire and the invention of the wheel.

When it comes to selfishness, however, how primitive do human beings appear!

## The co-creator went mad

The co-creator has created bombs – nuclear bombs, chemical weapons, biological warfare – which has brought him to the blasphemous position of facing the Creator saying: 'I have the power to liquidate life on our planet, indeed, I could destroy the earth (which has had the privilege of the redemptive incarnation of your divine Son) more than thirty times over'.

It is a matter of urgency to promote serious and profound studies in ecology.

# WE ARE IN GOD'S THOUGHTS THROUGH ALL ETERNITY

In God there is no succession of yesterday, today or tomorrow.

In him there is only *today*, always the same and always new.

In God there is no routine, no repetition. How tiring sameness is!

We have been called to life by God. And because in God there is only a perennial today, let's constantly keep this thought in our minds: always the same and always new!

If we had not been called to life, no one could have asked anything for us because we would not even have had a name of our own . . . This should be enough to drive us to an attitude of authentic humility. Pride is simply a lack of intelligence.

It was not only the saints and the prophets who were in God's thoughts through all eternity . . . It was not only the kings and the wise and great men of this world: the most neglected creatures, the people who live in subhuman conditions of misery and hunger are sons and daughters of God, called to life by the Creator.

God does not accept praise, gifts and honours from those who have no eye or heart for the human family, his sons and daughters of all races, all colours, all languages and creeds . . .

No one was created to be a slave or a beggar.

# WE ARE THE INTERPRETERS OF CREATION AND SINGERS OF GOD'S PRAISE

The Psalms teach us to lend our voice to all creatures: to the mountains and the waters; to the trees and the birds; to the light that comes from above and to the earth that provides for us; to the creatures of the sea, from the tiniest fish to the whale.

Who has seen the same dawn twice? Who has seen the same sunset twice?

It is a pity that there are people who will go through life never having thought of watching the sunrise! Or without thanking our dear friend at nightfall!

Ah, but would you like to have seen the splendour of the act of creation? Then just think, creation is made anew, instant by instant, at God's hands.

# WE LIVE IN GOD

Any time, day or night, at home or in the street, wherever we are, we live bathed in God. If we always kept this in mind, it would be impossible to sin.

How could we be filled with hatred, if we live night and day in God who is love?

How could we be filled with pride, if we carry with us night and day the Creator of all the worlds?

And to think that often we do not stop even for a moment to greet him, to thank him.

We can think about all our faults and weaknesses and about our every sin and we will come to the same conclusion: all we need for a life without sin is to remember, as the apostle Paul teaches us, that 'in him we live and move and have our being'.

24

# TEACHING THAT SEEMS EXCESSIVE

Christ teaches us to be perfect as our heavenly Father is perfect.

He wants to warn us against ever feeling perfect enough, holy enough . . . the sanctity of the Father is our goal. To live day and night, always and everywhere, in God. To think about God, as easily as breathing . . .

The problem is that, day and night, everywhere and always, we breathe (and we couldn't live without breathing), but how often we forget the air! Usually we only ever think about it when it is not there!

The same danger exists in our relationship with our God and father.

To be a saint does not mean never to sin. It means to start again with humility and joy after each fall.

# Humility in the incarnation and redemption

# GOD'S DAZZLING ANSWER

In creating, God revealed an enormous prefer-
ence for human beings. In us he summed up
several worlds.

> The minerals are our kin:
> we each occupy space and we are sensitive to
> the law of attraction.

> The plants are our kin:
> like them we are born, we are nourished, we
> grow and we die.

> The animals are our kin:
> sometimes we are surprised at the awakening
> of the animal within us.

> The angels are our kin:
> our body is the bearer of a spirit and is kin to
> the heavenly spirits.

We share in the very nature of God, in his intelligence and in his creative power.

We have been raised to the glory and the responsibility of co-creators.

We have characteristics that are specifically human, like our unmistakable way of smiling.

Yet man and woman felt themselves so great that they gave in to the temptation of imagining that they were just one step away from becoming God.

And the Creator could have crushed the human creature; he could have completely suppressed the human race on earth.

But on the contrary, God's answer was dazzling, divine.

The Son of God, without losing his divinity, became incarnate, that is to say, he received a body like ours and a spirit like ours in the most pure womb of Mary, through the work of the Holy Spirit.

The man-God passed through the earth doing good. He brought us a divine message. He created a Church to support us in our journey on earth. Above all he suffered and died for our sins!

This is what it means to repay evil with goodness. This is the divine answer to our ingratitude and our pride.

## A Dock Worker's Comment

A dock worker, who was so strong that he was like a human crane, once told me discreetly, almost confidentially: 'I hope God will forgive me for saying this, but when it came to carrying burdens, Christ was a weakling. He fell three times under the weight of the cross on the way to Calvary. That would be child's play for me.'

The mistake of our friend here was to think that the weight of Christ's cross was merely of its wood. But the weight that overwhelmed the Saviour was all the sins of humanity from every place and every age.

It was for the same reason that he sweated blood on the Mount of Olives and was humble enough to let himself be comforted by the ministrations of an angel.

# HE WENT WITH HUMILITY TO DO GOOD

In order to spread his message, Christ chose the Apostles, humble people, most of them fishermen, including Peter, whom Christ was to leave in his place . . .

The crowds around him were poor and Christ addressed himself specially to them. He said:

'The Spirit of the Lord is upon me,
because he has anointed me to preach good news to the poor.
He has sent me to proclaim release to the captives
and recovering of sight to the blind,
to set at liberty those who are oppressed,
and to proclaim the acceptable year of the Lord.'

He taught by the way he lived rather than by words.

When a group of men came to him, pushing a woman whom they had caught in the act of adultery, shouting and asking whether they should stone her or not, Christ was not in a hurry to answer. He was sitting surrounded by a crowd and then he started to write in the soil with his finger. One can easily imagine the curiosity of the accusers: 'What could Christ be writing on the ground?' Then, he raised his eyes and said: 'Whoever is without sin among you, let him cast the first stone.'

If one of us had said that, people would have come forward as if they had no sins. But before the eyes of the Son of God, no one dared.

When the accusers had gone away, Christ asked the woman: 'Where are your accusers? Does no one condemn you?' She replied: 'No,' and Jesus said: 'Neither do I. Go in peace and sin no more.'

Yet, we are always carrying stones in our pockets or our bags. How easily we judge, condemn and stone others!

Look in the new Testament and you will find many similar meetings that Jesus had, such as:
   with the Samaritan woman,
   with Mary Magdalen,
   with Zacchaeus,
   with Martha and Mary,
   with Peter.

# HE TRANSLATED HIS HIGH DOCTRINE INTO SIMPLE PARABLES

Christ taught his doctrine, for the benefit of the humble, through parables which are easy to remember and will never grow old . . .

One day he told the story of a man who had two sons. The younger asked his father for his part of the inheritance because he was bored at home and wanted to see something of the world.

The father was worried and he warned his son that he would suffer a great deal. He asked him what it was that did not satisfy him at home and promised he would do his best to make him happy.

But the son had made up his mind, so the father gave him the money. Giving him the money was easy, but watching his son leave home and knowing he would suffer was very difficult.

The young man went away. While he still had some money, he had plenty of 'friends' who took advantage of him. But as the money began to run out, the friends too started to disappear . . .

He had to find a job. That was not easy. Eventually he got work looking after pigs. But he was often so hungry that he was forced to eat the slops fed to them.

Then, one day, he was struck by the grace of God and he said to himself: 'I am stupid. Here am I, taking food from the pigs, while not even the servants in my father's house suffer what I am suffering. I'm going to go back home and throw myself on my knees in front of my father saying: "Father, I have been punished."'

(The whole tale of the prodigal son can be found in Luke 15:11–32.)

The same voice can be heard teaching in other parables, for instance:

The parable of the sower (Matt. 13:3–9; Mark 4:3–9; Luke 8:5–8):
Shouldn't we give thanks to the divine sower who doesn't wait until we become good before he sows?

The parable of the unmerciful servant (Matt. 18:23–35):
Don't we act like him at times?

The parable of the Pharisee and the publican (Luke 18:10–14):
Doesn't a Pharisee live inside us? Sometimes we may think he is gone, but he is only sleeping.

The parable of the talents (Matt. 25:14–30):
It would not be sincere to say we have no talents. But it would also be foolish to say we have the monopoly of talent.

# THE APPARENT FAILURE OF THE REDEMPTIVE INCARNATION

When evening came on Good Friday, the evening after Christ's crucifixion on Calvary, the failure of the redemptive incarnation seemed obvious. Nailed to the cross between two robbers, Christ died after three hours agony.

Apart from Mary his mother, John and Mary Magdalen, the only others present were the soldiers ordered to carry out the death sentence given by Pilate. Where were the Apostles? Where were these men who had spent three years living and working with Christ?

One of them betrayed and sold his master to those who wanted to kill him. Peter, who had been chosen as the rock on which Christ would build his Church, denied three times that he knew him, despite having been warned by the master that this would happen.

Deserted by his friends, nailed naked on a cross and receiving insults even from a fellow 'martyr', the bad thief, Christ died after three hours of utter agony.

And to make sure he was dead, a spear was thrust into his side.

He was buried.

Yet it is true that he had announced his passion and death, adding that after three days he would rise again.

**They would not believe**

And even his own disciples were unsure of his coming resurrection.

At dawn, Mary Magdalen went to the tomb where they had buried him. Carrying perfumes she was probably going to embalm the body of her master. The rock that blocked the entrance to the tomb had been removed. An angel who was standing there said: 'Christ whom you seek, has risen and he wants you to take this news to the Apostles.'

Peter himself thought the woman was imagining things. Nevertheless, he and John went to the tomb. They found everything as Mary Magdalen had told them. But they only believed in the resurrection after Christ himself had appeared to them, when they were gathered together in a locked room, afraid of the Roman soldiers.

Thomas, who was not there when Jesus appeared to the others, vowed he would only believe if Christ allowed him to touch the wounds in his hands, feet and side. Christ, with humility, accepted Thomas's challenge. Bearing in mind all those with a fragile faith, Christ invited Thomas to touch his wounds.

## The Emmaus Road

The Gospel of Luke tells the story of the disciples of Emmaus (Luke 24: 13–35).

They too had heard from Christ himself the prediction about his imprisonment, the torture he would have to endure and his death. And they also knew from Christ himself that, on the third day after his death, he would rise again.

They waited for two days. On the morning of the third day, their hope had practically disappeared and they left for Emmaus.

Certainly if we reflect on all the times when we have lost faith and hope so easily, we can appreciate the great patience of Christ with the disciples on the road to Emmaus.

A friend of mine who is a priest was meditating one day upon these same Emmaus disciples. He found it difficult to believe that after three years of living with Jesus they failed to recognize him when he joined them and began talking to them.

When there is a real friendship with someone, we are able to recognize them by the way they walk, their voice or even a cough.

My friend was meditating on this when there came a knock at the door. It was a poor person who wanted to tell his troubles to a priest.

The priest, immersed in his meditation upon the disciples of Emmaus, asked to be excused for not being able to listen to the poor man. He was, he said, really very busy. He even gave the man more money than he would have done normally, and went back to his meditation.

Soon afterwards it dawned on him that he had done exactly what the disciples at Emmaus had done: 'Christ knocked at my door. I saw him, I heard him, I talked to him and I failed to recognize him.'

# THE APPARENT FAILURE
# OF THE YEAR 2000

For a rose, a day is a long time . . .

A hundred years for a mountain must be like a single day for a human being.

We are two thousand years from the redemptive incarnation of Jesus Christ.

Incarnation? Yes. The Christian faith teaches that the Son of God, whilst always remaining the Son of God, by the work of the Holy Spirit took on the body of a man with a soul, just as every human creature does from the first moment of life in their mother's womb.

Redemptive incarnation? Yes. The Christian faith teaches that the Son of God became man, man-God, to take upon himself all human sin and to suffer, die and rise again for our salvation.

The two thousandth anniversary of the redemptive incarnation of Christ is a splendid opportunity to take stock of the situation, to see what we Christians have done, and what we are doing with the life, death and resurrection of our Lord Jesus Christ.

If we are honest, we have to admit that we have made very little use of it, if we apply Christ's own standard. He wanted us to build up a true human family with one Father, our Creator and God.

Christ summed up the whole of his teaching (he said: 'All the law and all the prophets') in two commandments:

To love God with all our hearts, all our mind and all our strength: 'This is the first and greatest commandment.'

Immediately though, he added: 'But there is another commandment, which is equal to the first: love your neighbour as yourself.'

Christ taught us that God, his divine father, is, and wants to be known as, the father of all human creatures, of all cultures, of all races, of all tongues, of all colours, of all creeds . . .

We can and we should consider ourselves as being brothers and sisters in blood, because the same blood that Christ shed for us, he also shed for other people.

46

On the eve of the year 2,000 the results achieved so far present us with a passionate challenge.

These statistics, gathered by the United Nations Organization should remain imprinted on our conscience:

*More than two thirds of the human race are living in sub-human conditions of hunger and destitution. More than two thirds of the children of God are living like animals.*

*20% of the human race consumes 80% of the earth's resources.*

*80% of the human race has to make do with only 20% of the same resources.*

These injustices, which are on an international scale, are found particularly in Third World countries. One example among many, is the case of land distribution in Brazil.

Brazil is an enormous country. It is nearly 40 times the size of countries like Britain or Italy.

And yet all this land is in the hands of only 8% of the population.

People have been killed for having the audacity to call for land reform, or for opposing violence and rejecting hatred as a solution to injustice.

# The Humility of the
# Holy Spirit

# HE BREATHED ON THE WATERS

When the wind blows gently or powerfully on a lake or river, it makes one think of the creative breath on the waters.

## The origin of the world

There are scientists who insist that the world began with an explosion. This explanation seems to satisfy them as if it were obvious that this is the whole story.

Normally scientists take great care before announcing a new discovery to the world, even with something like a new vaccine. They carry out rigorous experiments and ask other scientists to verify their findings. Only then do they announce their discovery.

Strange that once the 'Big Bang' theory had been formulated, it should so come to dominate the evidence. It would seem much simpler just to admit the notion of creation.

They are not believers. Paul VI liked to repeat that those who do not believe in God would be helped to overcome their unbelief, if we who have the joy and the responsibility of believing in our God and Father would give witness, not so much with our words but with our lives!

Let's ask the Holy Spirit to breathe especially upon those who do not have the happiness of believing in God.

## Creation never ceases

Disbelief in the Creator would be impossible if those who have doubts about the fact of creation had had the privilege of being present at the making of the universe.

And yet, the creation continues. How can one fail to recognize creation in the reproduction of plants, in the reproduction of animal life and of the human race?

There are various songs that speak of how creation has never ceased. Even if you have no great voice, you could learn some of them, and try to sing them. It doesn't matter if you only whisper them inwardly to yourself, but sing in praise of all creation, to the Creator, to God the Father and the Holy Spirit.

Do not fall silent at the proclamation that human beings can be in such a position: God wanted, and still wants, men and women to be his co-creators.

So we should not be alarmed by transplants in the human body, which will become more audacious and more reliable.

And we should not be surprised at how human life can be prolonged. Though what is still to be achieved is not the physical renewal of the body, but renewal within ourselves.

Special care is needed, however, when it comes to discoveries of the origins of life. They start with the borrowing of a womb, where the seeds of life from one couple find shelter in another woman who, for friendship's sake or for money, agrees to face pregnancy. But they end with the seeds of life flowering in test tubes.

It is true that a man cannot imagine what the nine months of pregnancy are like. But how much must a woman lose her sense of maternity if she entrusts it to another, or still more to a test tube!

The time may not be so far off when the greatest insult will be to call someone a test tube baby.

How wonderful is the coming together of the two seeds of life in a woman's womb! At that moment the Creator creates an immortal spirit and breathes it into the maternal womb. God does not prefabricate souls.

Parents can see themselves as God's partners; the Creator works together with them!

Yet no matter how close to us God chooses to be, he is God and we are his creatures.

When we reach our eternal home, having received the light of glory, we will be immersed in the divine mysteries. We will know for example, the role of God the Father, our Creator, and the role of the Holy Spirit with his creative breath.

We will be able to see the mystery of the Holy Trinity: one God in three distinct, yet equal persons.

# FAITH, HOPE AND LOVE

Human beings, within the limits of their own temperaments, are influenced by their families (if they have one) and by the environments in which they are born and where they live.

The Holy Spirit takes advantage of every opportunity to teach them with his inspiration how to live, how to do good and avoid evil.

And whoever receives more, has to respond more. When someone receives more, he or she is neither greater nor better than someone who has been given less. But all through life, even for those who receive more graces, every positive inspiration comes from the Holy Spirit. This is why when a priest recites the breviary, a prayer which is always made in the name of the whole of humanity, he starts by asking the Holy Spirit to open his lips: 'Open my lips, O Lord, and I will sing your praises.'

Divine grace is necessary before, during and after every act that is positive and holy.

For this reason, devotion to the Holy Spirit is important.

The more hectic life becomes, and the more numerous and stronger the attractions of the world, the more precious is the gift of faith, and the more it must be nourished.

Once, a woman, who wanted to live as a true Christian, complained to a priest about the many powerful temptations constantly assailing her. She even told the priest she had wanted to ask God to make her blind. The priest replied: 'Are temptations, however strong or numerous, really heavier and harder to cope with than living in God day and night, at home and at work, resting and relaxing, and lifting your awareness to the most Holy Trinity?'

## Thinking a little

In the present age, the more we suffer the lack of hope, the more the Holy Spirit has the task of bringing forth clear signs of hope.

Can you see signs of hope when faced with the ever greater absurdity of war in its various forms: nuclear, chemical, biological? Are they visible in the stockpiling of weapons?

Do you remember the statistics that show the crazy amount of money spent on armaments each year and think about all the good that could have been done with the same money?

Do you realize the ridiculous and blasphemous position of the arms race before the face of the Creator?

## Conventional Warfare

Do you realize the gravity of war, even without nuclear weapons, and the number of deaths it has already caused? What are we to think of those nations that make and sell conventional arms to Third World countries without resources enough to reduce the economic misery of millions of people?

Wars would come to a quicker end if arms were not donated to promote self-interest or sold for profit.

## The War of Poverty

This is a war that has already begun, and that has reduced more than two-thirds of the children of God who live on our planet to sub-human conditions of hunger and destitution.

## Ecological War

Human beings, who were raised by God to the glory and responsibility of co-creators, are destroying nature.

Capable of transforming barren wastes to fertile land, we seem proud of creating deserts.

It is not enough to have a clear vision of the various types of war. It is essential to discover the clear signs of hope made evident by the Holy Spirit.

# THE COURAGE
# OF THE MARTYRS

In the last thirty years there have been more martyrs than there were during the first centuries of the Church. The truth of this is now coming to light and will soon be published abroad.

People, young and old alike, are being murdered for the crime of encouraging, without hatred or violence, the promotion of human rights for the children of God living in destitution.

No one can say: 'Look at me! I am going to be a martyr!' To be called to the privilege and glory of offering one's life for Christ's message, for a more just and a more humane world where everyone feels they are brothers and sisters and sons and daughters of God, is a personal choice of God the Father.

No one should trust in their own strength, which is non-existent. Those who are chosen to the grace of martyrdom, need not worry about the superhuman courage needed. The Spirit of God will see to that.

What a wonderful light of hope: the martyrs, the numerous martyrs of today. And to think there were people who considered Christ incapable of creating martyrs any more!

# EVERYONE IS CALLED

When God our Father calls us to life, he chooses a human family into which we are to be born, a nation to which we are always linked during our time on earth, and respecting our freedom, he establishes a plan of life for each of us. He has thought of us from all eternity. For him there is no future and no past, but only today, always the same and always new.

The Holy Spirit sees to it that we live up to the Father's plan as well as possible.

It is wrong to think that the Father only has a plan for the life of a few creatures. None of us, no matter how humble, is thrown into life without a direction to follow, without a precise way ahead, left merely to chance . . .

There is divine Providence! Chance does not exist.

There are many people who pass through life without knowing that divine Providence has thought about each one of us, that Providence has a plan that can be upset by us or by others . . .

Vocation is a call – a call of God to the different states of life and to different professions.

Vocation to the priesthood or to the religious life certainly deserves special care and the prayers of us all. This is not because priests and religious are more important or greater than others, but because of the role they play in the Church and in humanity.

Yet have you ever thought how great is the call to matrimony? Always remember that fathers and mothers are partners with God in bearing and bringing up their children.

Have you ever thought that for every honest profession there is a call from God?

I know a priest who likes to shake hands with the dustmen when they are loading the refuse onto the lorry. They try to clean their hands on their clothes.

62

The priest, rightly, says: 'No work stains human hands. What makes hands dirty is stealing, or greed, or the blood of our neighbours!'

But would you like to give Christ a special joy? Then work for an increase in religious and priestly vocations: let them be numerous and holy.

There are those who say they have never felt a vocation, in the sense of a call from God. But God certainly calls. Only those who create silence within themselves will hear . . . the silence of hope, of generosity and of love.

# THE SPIRIT GIVES US EXAMPLES

The Holy Spirit takes great care in sending us people who become our models.

To all Christians of every place and time, the greatest example is of course Jesus Christ himself. But it is encouraging to find in different circumstances and at different times, people who imitate Christ and who become an inspiration for us. They are models that are always and universally relevant.

## Mary

We know that a creature like us, whilst remaining a creature, was chosen by our heavenly Father to be the mother of his divine Son, our mother, mother of mothers. And she says in the 'Magnificat', her song of praise, that she was chosen for her humility: God looked on the lowliness of his servant ... For that reason all generations will call her blessed.

The Father saw that amongst all women no one would be as humble as she when reaching the greatest glory: to be the mother of Christ, the mother of God.

She experienced the greatest joy that a human creature can experience. On the other hand, she also plumbed the depths of suffering . . .

The Son beyond compare cannot resist any request of the best of mothers.

## Joseph

He deserved to be chosen by the heavenly Father to receive the title, on earth, of being the father of Jesus Christ – even though, as we know, the birth of the child-God was through the Holy Spirit's miraculous intervention.

For his humility he received the glory of participating in the divine mysteries.

He probably died before seeing Jesus start on the mission for which he had come to the earth. Nevertheless, it must have been a wonderful way to die, with the care and the companionship of Mary and of Jesus himself, who called him father!

There is no doubt that the Son of God helped St Joseph in the carpenter's shop in Nazareth. Christ, who fully deserves the title 'Christ the King' – king of justice, love and peace – also deserves the title 'Christ the Worker'.

## The Angels

Certainly it is the Holy Spirit who stirs up in us a devotion to the angels of God. In the preface to the Eucharist we are asked to unite our voices to the voices of the angels. They are models of prayer, humility and discretion.

What is the name of your Guardian Angel? No one knows. While on our way to our Father's house, we could make up a name for him, so that our relationship with him becomes more real. What a joy it will be when you meet him for the first time in heaven and he tells you his real name, the one given to him by the heavenly Father!

# THE COURAGE OF THE OPPRESSED

When wealthy people go on holiday, the travel agent is careful to send them to places that are beautiful and interesting, rather than to places that may make them feel uncomfortable.

So, when the wealthy tourist sees the statistics published by the United Nations that 'more than two-thirds of the human race today live in sub-human conditions of poverty and hunger', they think it is a wild exaggeration or extreme left-wing propaganda.

Yet it is all too true, as can be witnessed in many so-called Third World countries, including Brazil.

Frequently when one visits certain destitute areas (often called by the United Nations areas of absolute poverty), one is shaken to the roots by the terrible conditions in which people live, and one says: 'This is not even fit for animals. But you, my friend, are a human being, a child of God. We are brothers and sisters because we have the same Father, our Creator and God.'

I am the first to admit that I have already proclaimed this terrible truth, ten, twenty, a hundred times, both in writing and in speech. And, with God's help, I will never tire of repeating it, as many others do, until by the grace of God we see justice obtained for all, decisively, with courage, but without hatred or violence.

The amazing thing, the surprising thing, thanks to the Holy Spirit who upholds and heartens the oppressed, is that to this very day I have never heard anyone saying: 'O Son of God, you've got everything; while we are merely Cinderella-like step children of God.'

We have no right to abuse this patience of our people. Faith and the Holy Spirit teach us to do much more than we have; we must go beyond just working 'for' the people and start working 'with' the people.

It is the Holy Spirit who guarantees the authenticity of the local church groups and it is he who enables even the uneducated to teach and to prove that no one is born to be a slave or a beggar.

The saying 'a united people will never be beaten' is thus realized and it takes on a truer, more Christian dimension:

*A people united and organized, a people united and relying on the grace of God, will rise up from poverty without hatred or violence, but with decision and courage.*

It is true that the tempter exists. He will do his utmost to create the maximum of confusion, but the divine Spirit will always give us his strength and his light.

# In Praise of Humility

# THOSE WHO DESPISE HUMILITY DO NOT KNOW WHAT IT IS

Today there are those, probably many, who despise two important virtues: obedience and humility. Here I want to look at humility, so that we can understand God's humility better, though obedience too has a special place in human life.

The truth is that those who despise humility do not know what it is. Far from being a virtue of slavery, humility is a synonym of truth, particularly truth about the virtues we have or think we have.

A person can try to grow in humility by seeking, truthfully, to see what qualities they have or have not, and in the case of having some good points, they can see how much is there naturally, how much has been given, how much has been acquired by effort. But after that, they should realize that every good thing in a human creature has come from God.

For humility is not false modesty. An intelligent person, for example, cannot pretend they have no intelligence, but should be aware that it is a gift from God and not think that they are as bright as if the sun shone in their heads, whilst others have brains full of breadcrumbs.

God neither forgets nor denies his infinite intelligence. The humility of God consists of living with human limitations and human meanness, just as a father adjusts his pace when walking with his small son who is not yet able to cope with long strides.

Why don't we study the grandeur and beauty of true humility?

# WAYS OF PRAISING THE HUMILITY OF THE FATHER

If you are the father of small children, try to be with them at least for a while on Sundays, and say, even if they find it hard to understand: 'You call me father, and so I am, but we all have a Father who is in heaven.'

For a moment you can draw with your own words a picture of the greatness and the goodness of our heavenly Father. You could end by saying, in your own way: 'Ask our Father in heaven that all fathers on earth, and I too, should try to deserve the name of father by trying to be good like him.'

Why not pray the Our Father while holding hands with your children?

When your children grow up, you could then try to recall, at least once a year, perhaps on a birthday, how you used to pray with them every Sunday. You could do this even with your married children, and say: 'Let's ask God that

if you become parents, you might be always more like him.'

Take your children, whenever you can, to see the sunrise or the sunset. Show them the beautiful things where you live: the mountains, the lakes, the huge trees, the flowers in their glory . . .

The Father, the creator of so much beauty, has the humility to go out and meet so many people who never have time or can never be bothered to gaze, even for an instant, on the wonders he has made!

Be an example of how to respect Sunday as the day of the Lord. Certainly nothing is better than to go to the Eucharist, which is celebrated by Jesus Christ himself. It is the Son, the Beloved, who helps us to glorify our Creator and Father, the Father of all, the Father of fathers!

From what is said here, it may seem that glorifying the humility of the Father is only a problem for fathers. The fact is that mothers need hardly need any suggestions to help them. They tend to let their imagination work and their hearts be moved as they come up with beautiful suggestions as to how to lead their children, and all the members of their family, to praise the humility of the Father and to sing the glory of the Lord's day!

# WAYS OF PRAISING THE HUMILITY OF THE SON

To honour the humility of the Son, there is nothing better, and greater, than both to be a living example and to teach with words. In the Eucharistic consecration we can see that the celebrant is always Jesus Christ, no matter who the priest (or priests) may be.

### Christ and children

Christ loves children!

On Sundays, Christmas, Easter, Ascension Day, try to get together with a group of children to play and sing with them, not forgetting to give them something good to eat, as an act of praise to the peerless friend of children.

### Christ and the poor

Jesus Christ identified himself with those who suffer.

Today in our country* one can find poverty and suffering in every developing town, although with different numbers involved and at different stages of degradation.

You could go and visit these sub-human places, even taking friends or relations with you, not out of curiosity but to meet Christ who becomes one with those who suffer.

Do not be alarmed if you find Jesus Christ not in a cot but in a wooden box from the market.

Do not be alarmed if you find mindless children.

Ideally you would not start working in isolation, but with some project that has already begun, in a parish for instance.

* Brazil. The fundamental thought expressed in this passage, however, is meant to have universal application. [Editor's note]

# WAYS OF PRAISING THE HUMILITY OF THE HOLY SPIRIT

Do you have any local celebrations in honour of the Holy Spirit? You could take part in them. Perhaps there are some festivals that could be revived. You could talk to the local clergy to see whether there is anything being done, or that could be done, to honour the Holy Spirit.

Who is guiding the Church of Christ to seek unity, so that Catholics are looking at their brothers and sisters of the Reform, and discovering what is common more than what divides?

Who is guiding the Church of Christ to be united in prayer and work for the poor, striving to create a just and more humane world, even together with those of other religions who believe in our Creator and Father?

Who is working for peace in the world?

Who inspires and feeds non-violent activity?

Who is behind the growing urge to unite the world which is divided into the First World, the Second World, the Third World, the Fourth World?

Who is helping us to go back again to the desire of our Creator and Father that humanity should be a single family, with one Father for all and all being brothers and sisters not merely in name but in fact?

# II

# The School of the Universe

# EARTH, SISTER EARTH

Teach us
To continue the Creation
To help the seeds
To multiply,
Giving food
For the people
And for the beasts.

Teach us
To further the joy
You never tire of offering
When weary travellers find you,
A signpost to their home.

Teach us
To make the horizon
Become a beautiful image
Of Creation's grandeur.

Teach us
To accept
The mediation of those
Who wish to unite us
To our fellows,
As you accept the gift
Of the water that binds
Land to land,
No matter how great
The distances!

What do you suffer
In the dust of deserts?
How do you look upon
Those of us who,
Though capable of transforming
The waste to lushness,
Prefer to be creators
Of barrenness?

And how do you rejoice
In the rain
That brings forth your fruits?
And what pain do you feel
At the storms
That drown you with floods,
Destroying plantations,
Crushing houses and the lives
Of animals, of plants, of people?

How great is the lesson
You give us,
O Earth,
More than sister:
Our mother Earth!

All our lives
We walk carelessly across you,
And when life leaves us,
With no shadow of resentment,
You open up to us
Your maternal bosom
To keep
Our flesh,
Our ashes,
For the joy
Of the resurrection.

# BROTHER FIRE

Are you aware
Of your beauty?
Are you aware
Of the artless grace
With which you fulfil
Your high task
To conquer the dark?

Who taught
Your flames to leap,
O dancer's envy,
Light of step,
Spinning in long pirouettes
None can forget?

Often you have the pain
To offer human life
As a sacrifice through
Your touch of flame.

How do you find the courage
At such times, as you rise
In impetuous strength,
To give battle to the water;
And what are your thoughts
Of the humble-noble fire-fighter?

With simplicity
You feed
The wood stove
And prepare the modest meals of
The families of the poor.

Do you not tremble with horror
When they speak of you
As a sign of punishment?
Yet with what great respect
Do you consume and transfigure
The bodies of the martyrs!

Thank you brother Fire,
For the warmth you bring
To freezing bodies
Which have only you
As their help and salvation.

Thank you brother Fire
For teaching us
To have a warm heart,
And helping us avoid
The emptiness of
A heart cold as stone.

Thank you brother Fire
For the song in your flames
When, overcoming the darkness,
You become praise most pure
For our Father and Creator.

# WATER, MY SISTER WATER

When you were created
Did you yet know
How many would be
The things you must do,
From the most noble
And beautiful
To the most base
And desolate?

Yes, you are beautiful
In the stillness of lakes,
In the flowing of rivers
(As humble brooks
Or as the rushing of rapids),
In glittering cascades,
In the oceans which leave in us
The lingering images of
The Infinite.

Yet for those who have eyes to see
And ears to hear,
You are still more beautiful
As you labour, with joy,
On your round of lowly tasks:
The washing of clothes,
The cleaning of floors,
The quenching of imperious thirst.

And impressive are you
In your ceaseless travel,
Lifted from the earth to the clouds,
And coming down again from heaven,
To bring life
To the plants,
To the animals,
Life to the human race.

How did you receive
The dreadful commission
To bring the flood
And storms at sea
And wild lashing of the tempest?

Do you know
You give the chance
Of heaven as the reward
Of those who offer you
To quench the thirst
Of their brother or sister?

Do you know
The sins of the human race
Are to blame
For the many pains
You are forced to cure?

Thank you,
Sister Water.

Forgive
Us who make you
Perform cruel tasks.

Thank you,
Above all
because you help us
To praise
The Creator and Father.

# BROTHER BIRDS

Do you think much
Of the beauty of your plumes,
The enchantment of your song,
The grace of your flight?
Are you free
From pride and vanity,
Or must you too
Defend yourselves from
These failings that
Humiliate us so much?

What joy you must feel
At the first flutterings of
Your little fledgelings,
And how you must remember
Their first attempts to sing.

You surely must have thrilled
At soaring up, up in the air,
And racing to the horizon –
Flying, flying, flying,

Until you drop with exhaustion,
With the ripeness of emotion.

What do you suffer
When you are locked in cages;
How can you sing
When you still have wings
To prove you were born
For freedom?

What do you think
Of a human voice in its beauty?
Is it anywhere near
The sweetness of your song?

When one of you dies,
Is there sorrow?
Or do you believe
That you too
Will one day rise up again?

Lord, in the name of those
Who have no voice for singing
I offer you
The most beautiful songs
Of your birds.
And I ask of you

That we
May feel ashamed
Of creating prisons for them.

Prisons for those
Who received from you
The mission to fly.
Let us be ashamed
Of listening to the singing
Of caged birds,
When a song needs freedom more
Than it does wings!

# BROTHER AIR

Have you noticed
How we humans
For all our civilization
Transform you,
The guarantor of life
Into a spreader of poisons?
Have you noticed
How we humans,
Gifted with intelligence and reason,
Are wounding nature
And preparing disasters
For ourselves
With our own hands?

How do you feel
When a thousand sounds,
Words and music,
Threats and songs of love,
Pass through you each instant?
How marvellous it would be
If everything you carried
Were at the service
Of peace and goodness.

Do you feel the difference
Between the flight of a bird
And the throb of an aircraft?
Do you see the spaceships
Shooting by?
Do you go with them?
In those far reaches
Is there still need of you?

No doubt you have knowledge
You cannot reveal . . .
No doubt you know well
The things beyond earth,
And always and everywhere
You feel in a wonderful way
The presence of your
Creator and Father.

Do you know that
You give us
Of all images
One most close to God?

We live
Inside God
Everywhere
At all times,
Just as we live
Inside you.

And just as we
Think rarely of you
By day and still
Less at night,
So do we rarely
Think of our God.

If we are without you
Even a few seconds,
Life is unbearable!
Only then do we think
Of you.
Our God fades from our thought
Even more, yet he never
Complains of the way
We leave him alone;
And he never leaves us!
O Air, please teach us
To think of you
And still more, much more,
To think of the Father,
Our Creator and yours.

# TREES, SISTER TREES

Ah, trees,
Do you live in peace,
In harmony,
Despite the differences
Between you
Which to the human eye
Seem so immense?
How do you feel,
You towering palms,
You massive oaks,
You giant baobabs,
As you stare down upon
A tiny bush?
Are the fruit bearers,
Apple, mango and coconut,
Tempted to mock
Those who have simply
Leaves and thorns?

Do you welcome
The birds' nests
And children who swing
From your branches?
What is it like
When the leaves fall
And your branches are draped
In snow?
Does Spring give a hint
Of the thought of resurrection?
And what is it like when
Drought wounds the land
And you trees seem to raise
Up your arms
In silent prayer
To our Lord and Father?

What do you feel
At the plucking of your fruits?
Is the stoning
The worst part?
And what is your pain
When your branches are pruned,
Or when a tree entire
Goes to the timber yard
To be cut
Into house or bridges or chairs or beds?
Do you understand the purpose
Of a bridge
And the vivid symbol it is?
Do you realize your importance in
The building of a home?

Chairs and beds
Call to mind rest
And make us think
Of the family:
But what does it mean to you?

When lightning strikes a tree,
Is it true that the tree
Prefers itself to be struck
Than to see a person or house
Destroyed?

Remember, sister Tree,
That the Son of God
In order to reconcile us
With our Father and his,
Bore a heavy cross.
Three times he fell
Under that burden,
And wanted to die
Nailed to the cross
In order to save us.

# Meditations

## by

## Chiara Lubich

Chiara Lubich's *Meditations* are fast becoming considered by many people to be spiritual classics. They have something in them for everyone.

Yet a strange quality of darkness and light characterizes them. In a sense they dazzle us. Their beauty vibrates with such a depth of meaning that they seem almost from another world. And in a way they are. They express, very clearly, something of heaven's viewpoint. And the human and the heavenly way of seeing things rarely coincide.

Nevertheless, the appeal of the *Meditations* is hard to deny. They tug at the heart and fill the reader with a thirst for the Infinite, while at the same time pointing out quite down to earth ways of living. For these writings are nothing if not eminently practical, something that can be transformed into daily life, and indeed, when lived, they transform daily life with their own beauty.

Published by New City
ISBN 0 904287 29 7
Price £3.95